My First Book About OUR AMAZING Earth

Donald M. Silver & Patricia J. Wynne

DOVER PUBLICATIONS, INC.
Mineola, New York

To the many people who love our Planet Earth
—*Patricia*

For Ilona Royce Smithkin
Artist, teacher, chanteuse, mentor—and still amazing at 98!
—*Donald*

Earth, our planet, is a wondrous place. It's the third planet from the Sun, the star without which we could not survive. In this beautifully illustrated book, you will find answers to questions such as: What is Earth made of? What is a fossil, and how is it created? What causes volcanoes and earthquakes? How are mountains like Mount Everest "built"? Why is coal called a "fossil fuel"? You will also learn about some of the national parks in the United States, such as Grand Teton, Shenandoah, and Adirondacks, as well as Devils Tower National Monument. Easy-to-understand captions provide fascinating facts about lava, rocks, the continents, and much more. Plus, you can color each of the realistic illustrations with colored pencils, crayons, or markers.

Copyright
Copyright © 2019 by Dover Publications, Inc.
All rights reserved.

Bibliographical Note
My First Book About Our Amazing Earth is a new work,
first published by Dover Publications, Inc., in 2019.

International Standard Book Number
ISBN-13: 978-0-486-83306-4
ISBN-10: 0-486-83306-2

Manufactured in the United States by LSC Communications
83306202   2019
www.doverpublications.com

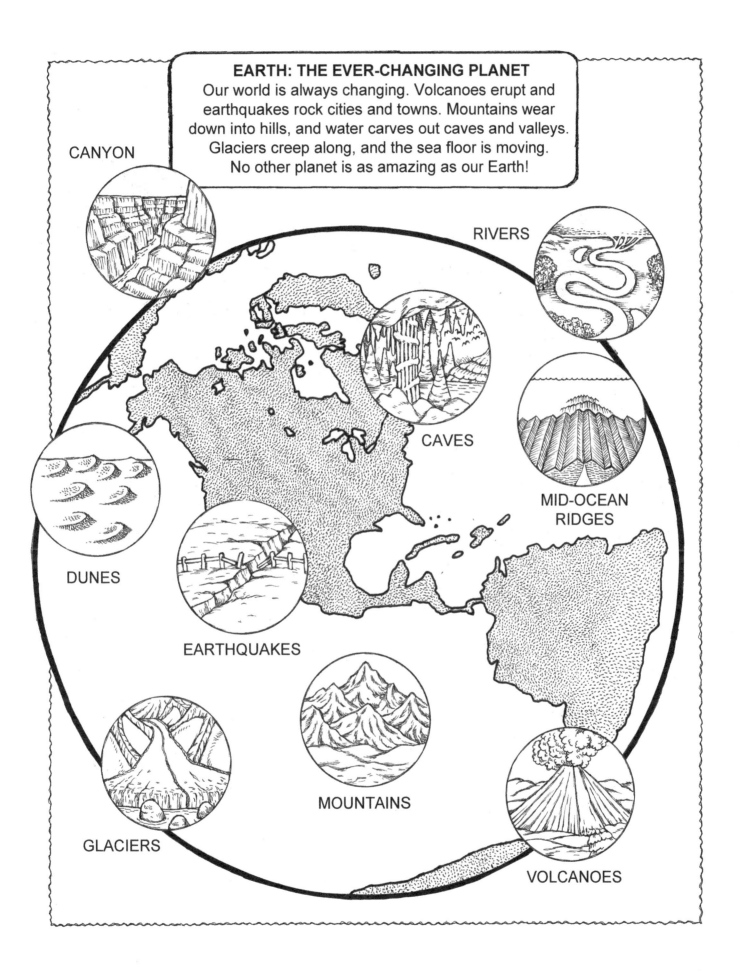

EARTH: THE EVER-CHANGING PLANET
Our world is always changing. Volcanoes erupt and earthquakes rock cities and towns. Mountains wear down into hills, and water carves out caves and valleys. Glaciers creep along, and the sea floor is moving. No other planet is as amazing as our Earth!

CANYON

RIVERS

CAVES

MID-OCEAN RIDGES

DUNES

EARTHQUAKES

MOUNTAINS

GLACIERS

VOLCANOES

1

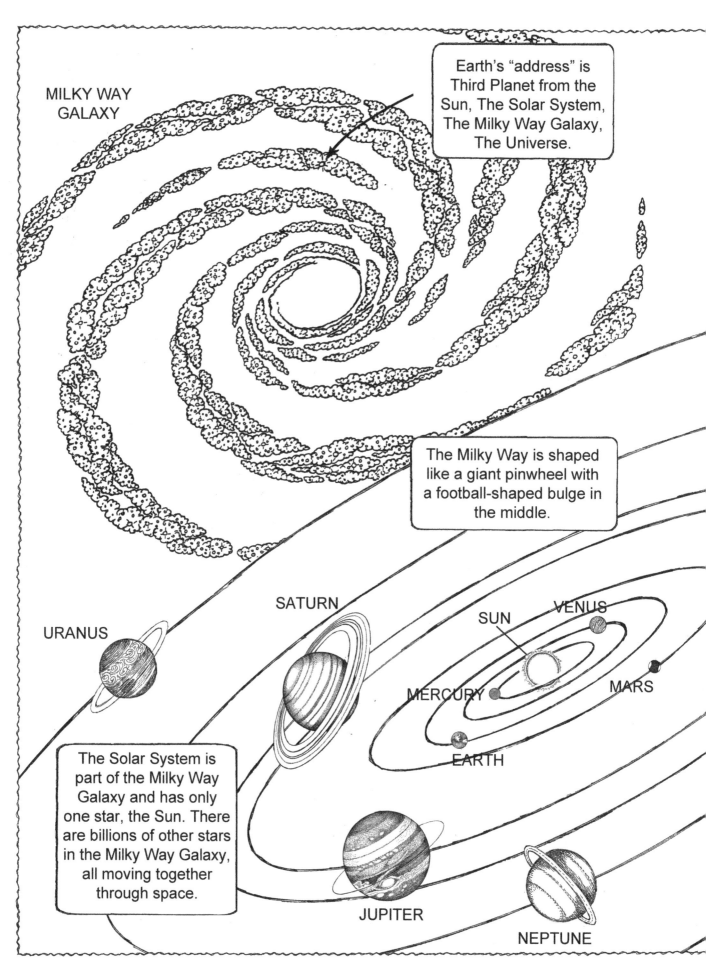

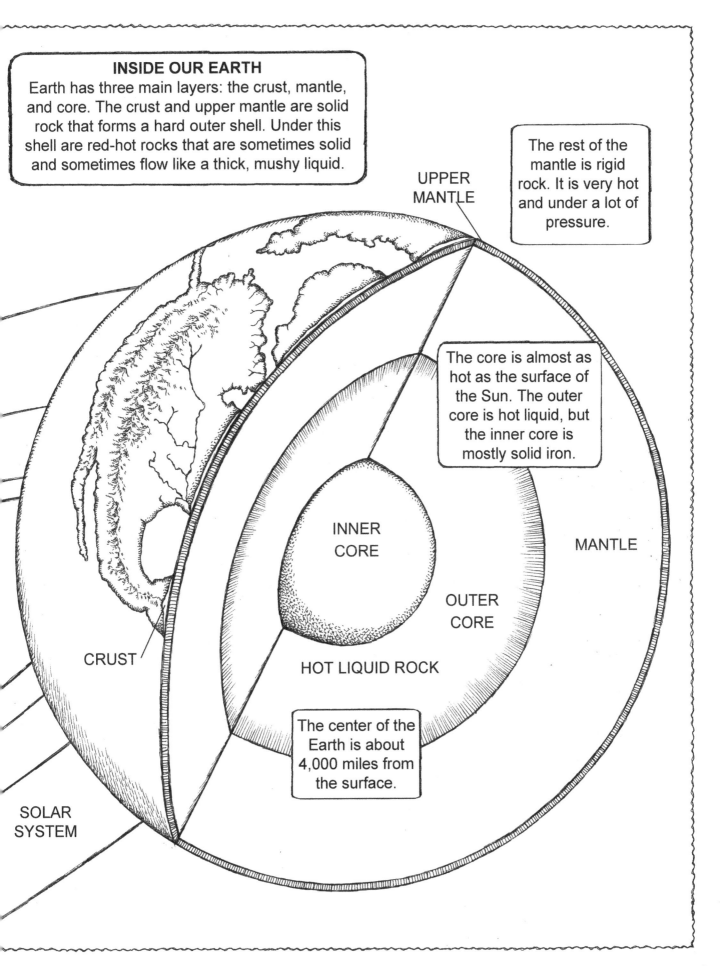

INSIDE OUR EARTH
Earth has three main layers: the crust, mantle, and core. The crust and upper mantle are solid rock that forms a hard outer shell. Under this shell are red-hot rocks that are sometimes solid and sometimes flow like a thick, mushy liquid.

The rest of the mantle is rigid rock. It is very hot and under a lot of pressure.

The core is almost as hot as the surface of the Sun. The outer core is hot liquid, but the inner core is mostly solid iron.

UPPER MANTLE

INNER CORE

OUTER CORE

MANTLE

CRUST

HOT LIQUID ROCK

The center of the Earth is about 4,000 miles from the surface.

SOLAR SYSTEM

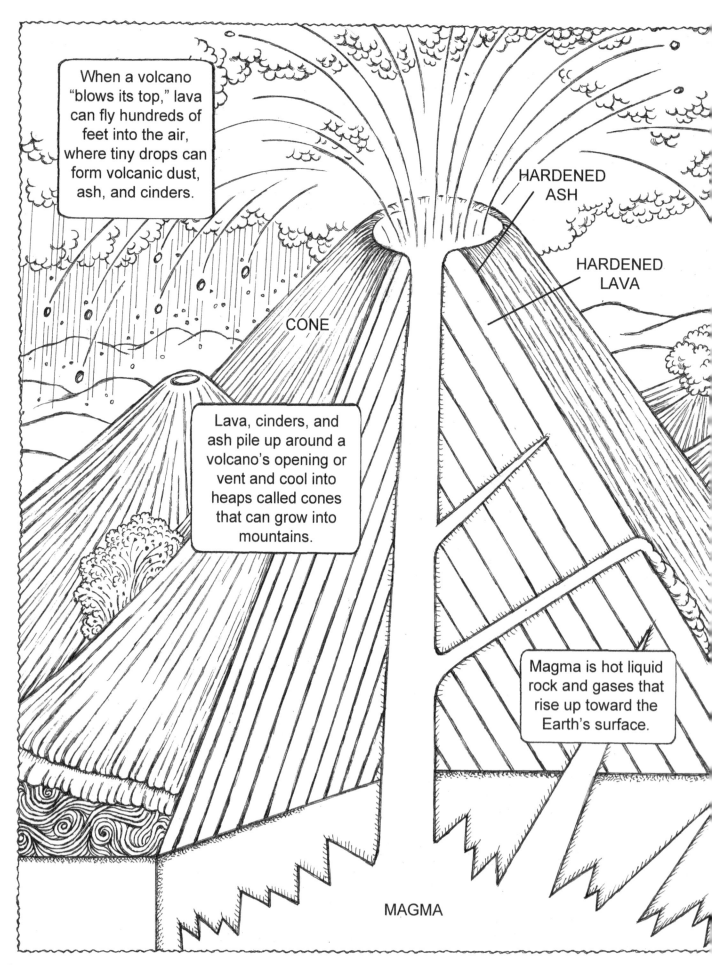

When a volcano "blows its top," lava can fly hundreds of feet into the air, where tiny drops can form volcanic dust, ash, and cinders.

HARDENED ASH

HARDENED LAVA

CONE

Lava, cinders, and ash pile up around a volcano's opening or vent and cool into heaps called cones that can grow into mountains.

Magma is hot liquid rock and gases that rise up toward the Earth's surface.

MAGMA

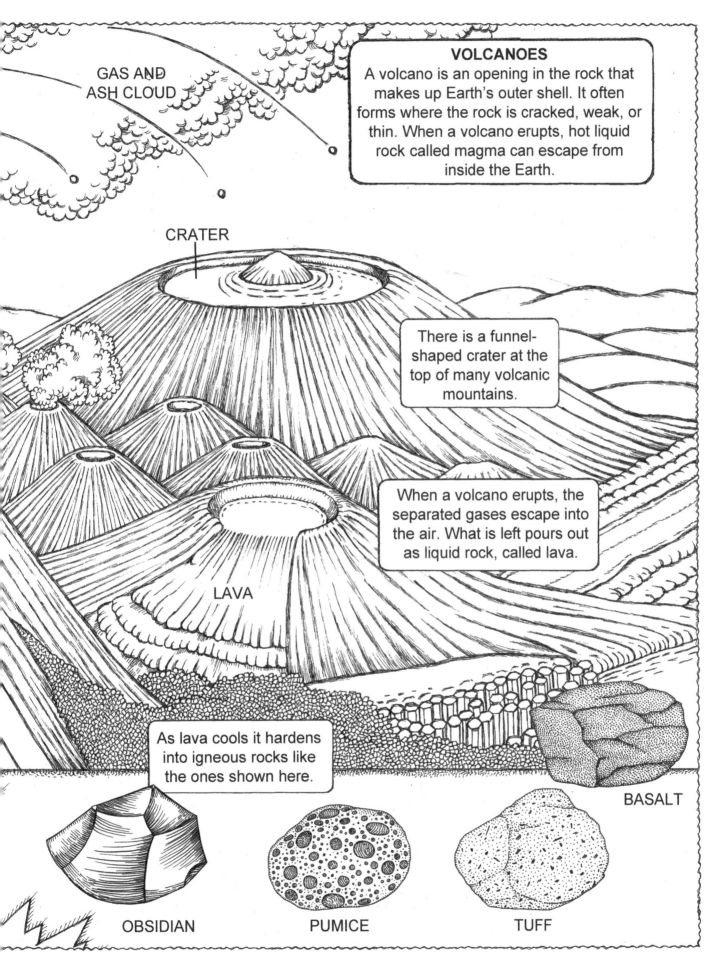

GAS AND ASH CLOUD

VOLCANOES
A volcano is an opening in the rock that makes up Earth's outer shell. It often forms where the rock is cracked, weak, or thin. When a volcano erupts, hot liquid rock called magma can escape from inside the Earth.

CRATER

There is a funnel-shaped crater at the top of many volcanic mountains.

When a volcano erupts, the separated gases escape into the air. What is left pours out as liquid rock, called lava.

LAVA

As lava cools it hardens into igneous rocks like the ones shown here.

BASALT

OBSIDIAN

PUMICE

TUFF

A VOLCANO AWAKES

In 1980 all was quiet around
Mount St. Helens
in Washington state.

1 Suddenly the mountain began to shake. Mount St. Helens had awakened from its 123-year sleep and was about to erupt. The north side exploded with more power than hundreds of atomic bombs.

2 A giant cloud of hot gases and ash rose from the volcano.

3 The cloud swept down the mountain, moved through a forest, and killed nearly every living thing in its way.

4 Lava and thick mud flowed down the mountain. The explosion lowered the mountain by 1,200 feet. After a few months Mount St. Helens fell back to sleep.

OUR NATIONAL PARKS

North Cascades National Park
Mount Rainier National Park
Crater Lake National Park

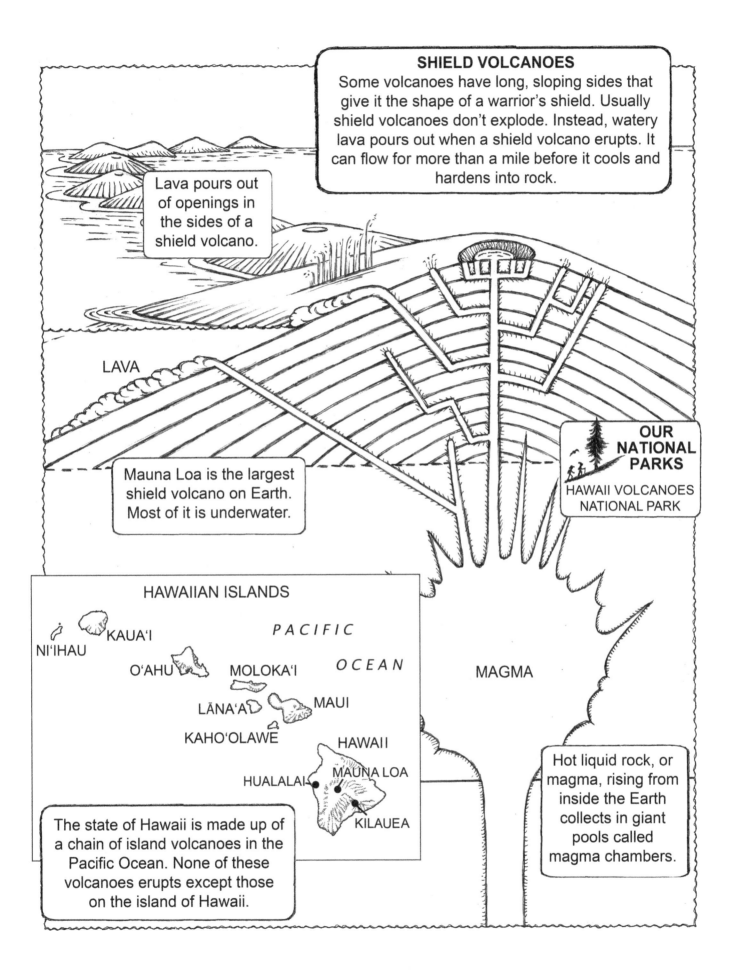

SHIELD VOLCANOES
Some volcanoes have long, sloping sides that give it the shape of a warrior's shield. Usually shield volcanoes don't explode. Instead, watery lava pours out when a shield volcano erupts. It can flow for more than a mile before it cools and hardens into rock.

Lava pours out of openings in the sides of a shield volcano.

LAVA

Mauna Loa is the largest shield volcano on Earth. Most of it is underwater.

OUR NATIONAL PARKS
HAWAII VOLCANOES NATIONAL PARK

HAWAIIAN ISLANDS

NI'IHAU
KAUA'I
O'AHU
MOLOKA'I
LĀNA'A
MAUI
KAHO'OLAWE
HUALALAI
MAUNA LOA
KILAUEA
HAWAII

PACIFIC OCEAN

MAGMA

The state of Hawaii is made up of a chain of island volcanoes in the Pacific Ocean. None of these volcanoes erupts except those on the island of Hawaii.

Hot liquid rock, or magma, rising from inside the Earth collects in giant pools called magma chambers.

7

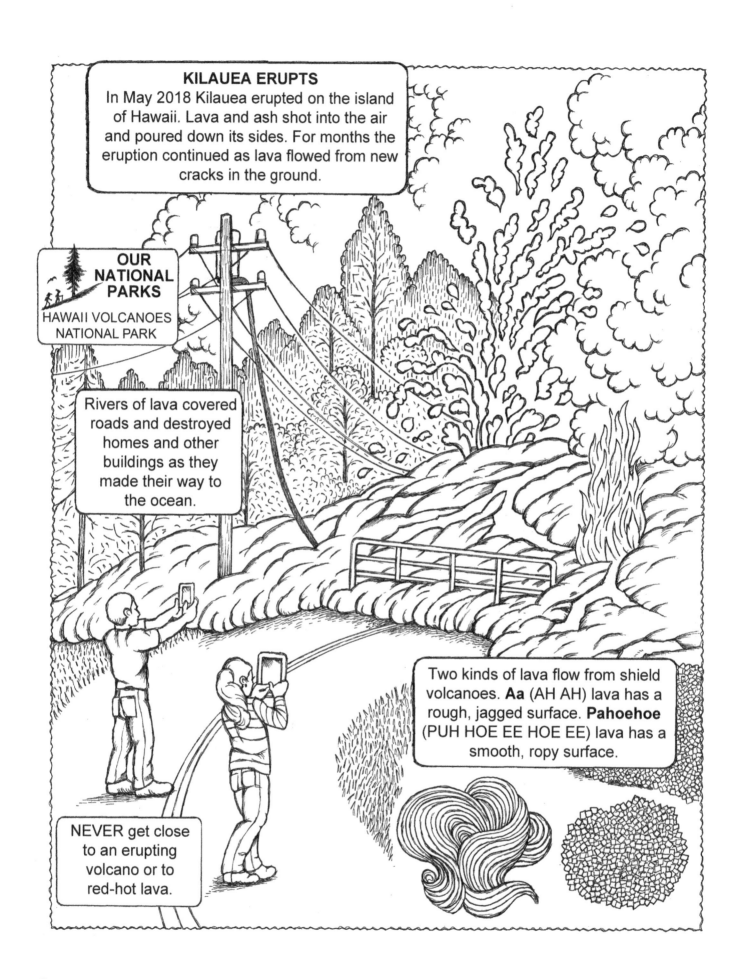

KILAUEA ERUPTS
In May 2018 Kilauea erupted on the island of Hawaii. Lava and ash shot into the air and poured down its sides. For months the eruption continued as lava flowed from new cracks in the ground.

OUR NATIONAL PARKS
HAWAII VOLCANOES NATIONAL PARK

Rivers of lava covered roads and destroyed homes and other buildings as they made their way to the ocean.

Two kinds of lava flow from shield volcanoes. **Aa** (AH AH) lava has a rough, jagged surface. **Pahoehoe** (PUH HOE EE HOE EE) lava has a smooth, ropy surface.

NEVER get close to an erupting volcano or to red-hot lava.

9

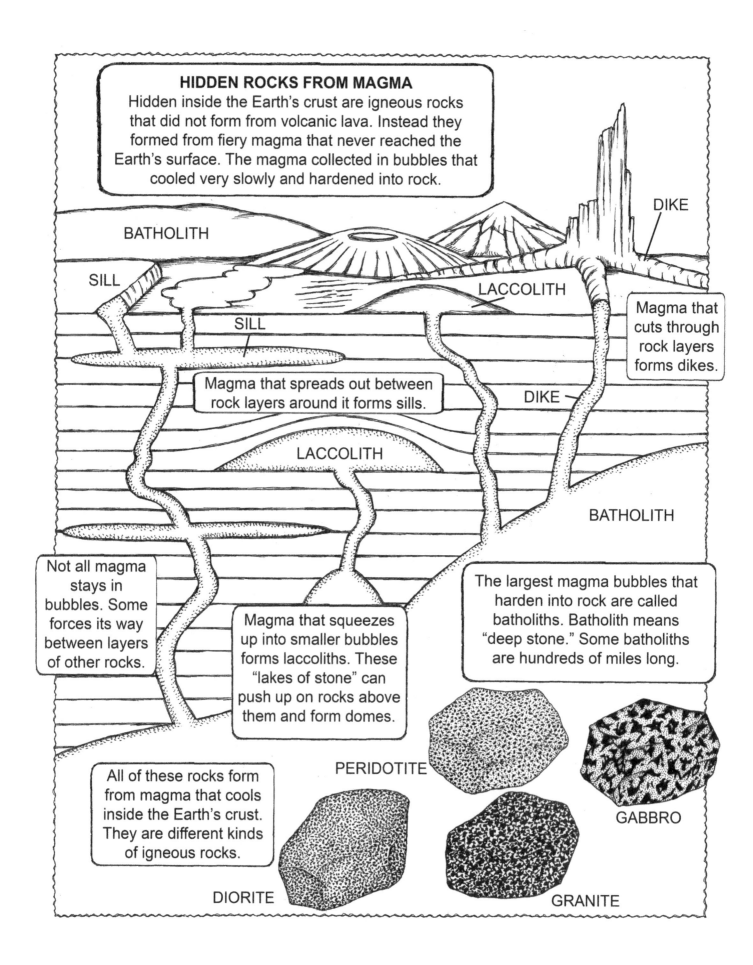

HIDDEN ROCKS FROM MAGMA

Hidden inside the Earth's crust are igneous rocks that did not form from volcanic lava. Instead they formed from fiery magma that never reached the Earth's surface. The magma collected in bubbles that cooled very slowly and hardened into rock.

DIKE

BATHOLITH

SILL

SILL

LACCOLITH

Magma that cuts through rock layers forms dikes.

Magma that spreads out between rock layers around it forms sills.

DIKE

LACCOLITH

BATHOLITH

Not all magma stays in bubbles. Some forces its way between layers of other rocks.

Magma that squeezes up into smaller bubbles forms laccoliths. These "lakes of stone" can push up on rocks above them and form domes.

The largest magma bubbles that harden into rock are called batholiths. Batholith means "deep stone." Some batholiths are hundreds of miles long.

PERIDOTITE

GABBRO

All of these rocks form from magma that cools inside the Earth's crust. They are different kinds of igneous rocks.

DIORITE

GRANITE

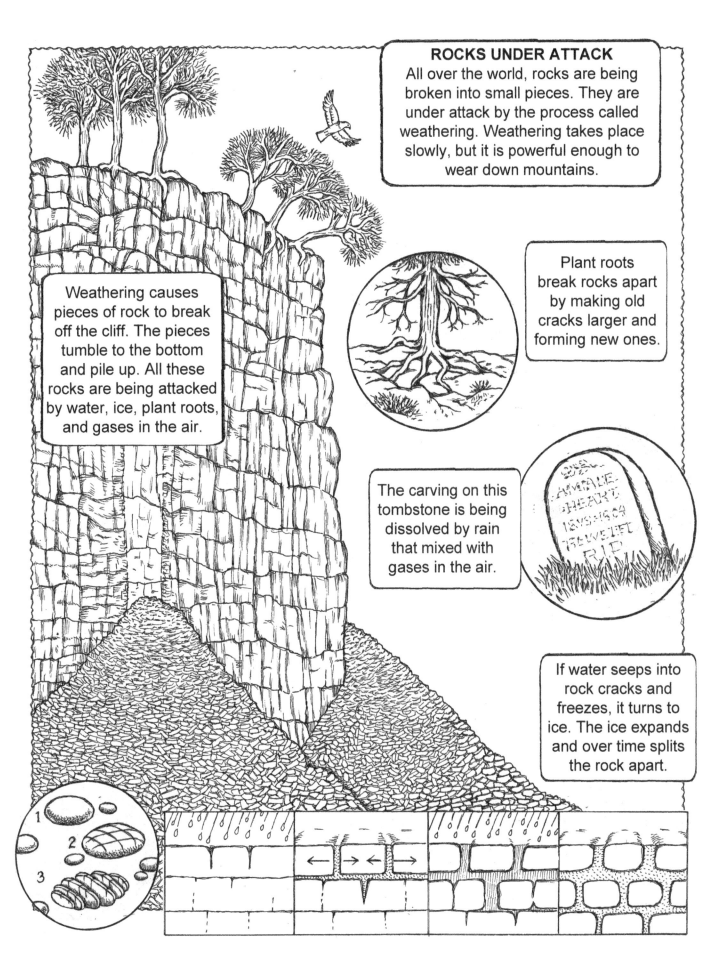

ROCKS UNDER ATTACK
All over the world, rocks are being broken into small pieces. They are under attack by the process called weathering. Weathering takes place slowly, but it is powerful enough to wear down mountains.

Weathering causes pieces of rock to break off the cliff. The pieces tumble to the bottom and pile up. All these rocks are being attacked by water, ice, plant roots, and gases in the air.

Plant roots break rocks apart by making old cracks larger and forming new ones.

The carving on this tombstone is being dissolved by rain that mixed with gases in the air.

If water seeps into rock cracks and freezes, it turns to ice. The ice expands and over time splits the rock apart.

ROCKFALL!
The famous cliff Half Dome rises 4,737 feet in Yosemite National Park. The cliff's strong granite rock hardened from magma and was once hidden underground. Like all rocks, Half Dome is slowly being broken down by weathering.

OUR NATIONAL PARKS
YOSEMITE NATIONAL PARK

As weathering loosens rocks on Half Dome, they fall down the face of the cliff. Thousands of tons of rocks can fall over several days, flattening trees and creating a vast dust cloud.

This granite rock at the top is covered in breaks. As the rock weathers, layers peel away from the rock. This peeling process is called exfoliation.

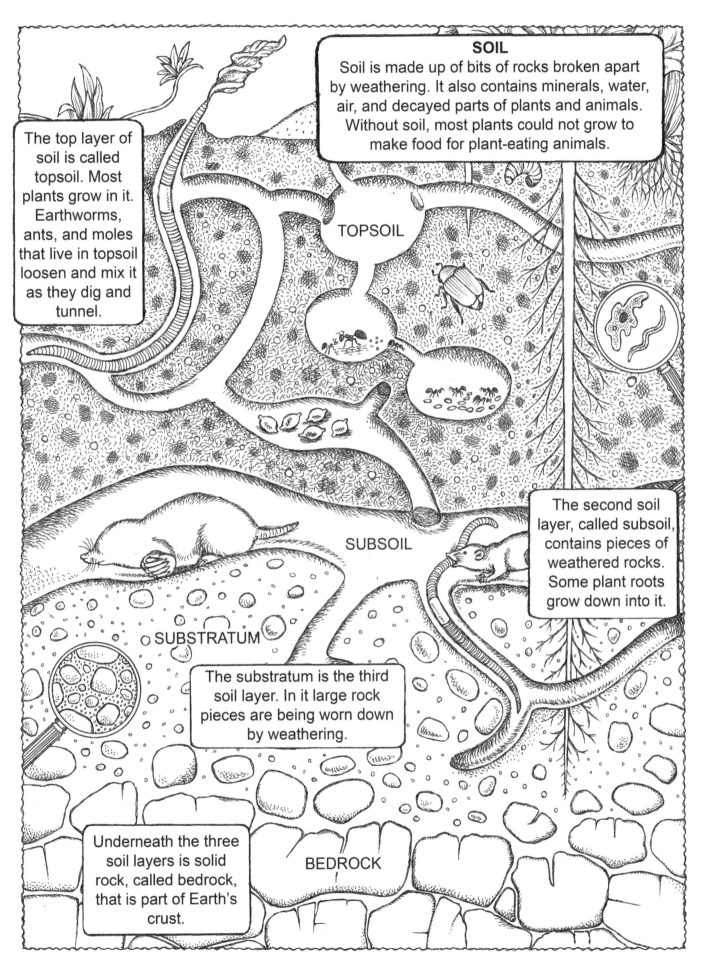

SOIL
Soil is made up of bits of rocks broken apart by weathering. It also contains minerals, water, air, and decayed parts of plants and animals. Without soil, most plants could not grow to make food for plant-eating animals.

The top layer of soil is called topsoil. Most plants grow in it. Earthworms, ants, and moles that live in topsoil loosen and mix it as they dig and tunnel.

TOPSOIL

The second soil layer, called subsoil, contains pieces of weathered rocks. Some plant roots grow down into it.

SUBSOIL

SUBSTRATUM

The substratum is the third soil layer. In it large rock pieces are being worn down by weathering.

Underneath the three soil layers is solid rock, called bedrock, that is part of Earth's crust.

BEDROCK

LAYERED ROCKS
Rain and melting snow wash sand, pebbles, and bits and pieces of other weathered rocks into rivers. Rivers deliver them into the ocean, where they sink in layers on top of each other. Each new layer presses on the layers below. Over time the lower layers are pressed together into layered rocks.

Conglomerate forms from rounded pebbles that are cemented together.

All the different kinds of weathered rock form sediment. Layered rocks are also called sedimentary rocks.

Shelly limestone forms with shells in it.

Limestone forms mostly from minerals in shell pieces and pieces of coral.

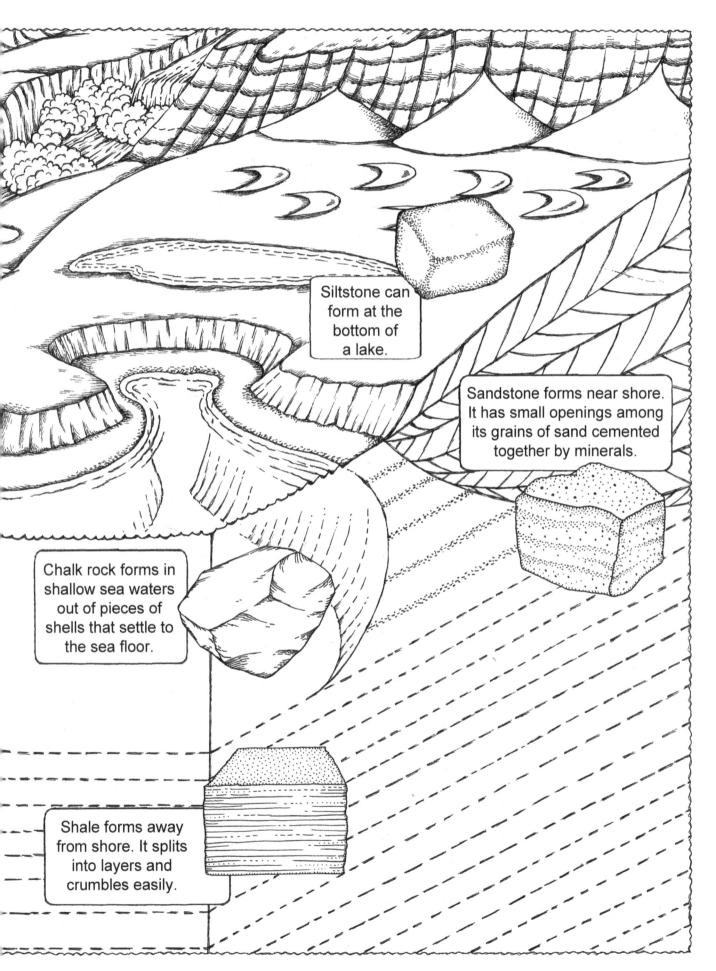

Siltstone can form at the bottom of a lake.

Sandstone forms near shore. It has small openings among its grains of sand cemented together by minerals.

Chalk rock forms in shallow sea waters out of pieces of shells that settle to the sea floor.

Shale forms away from shore. It splits into layers and crumbles easily.

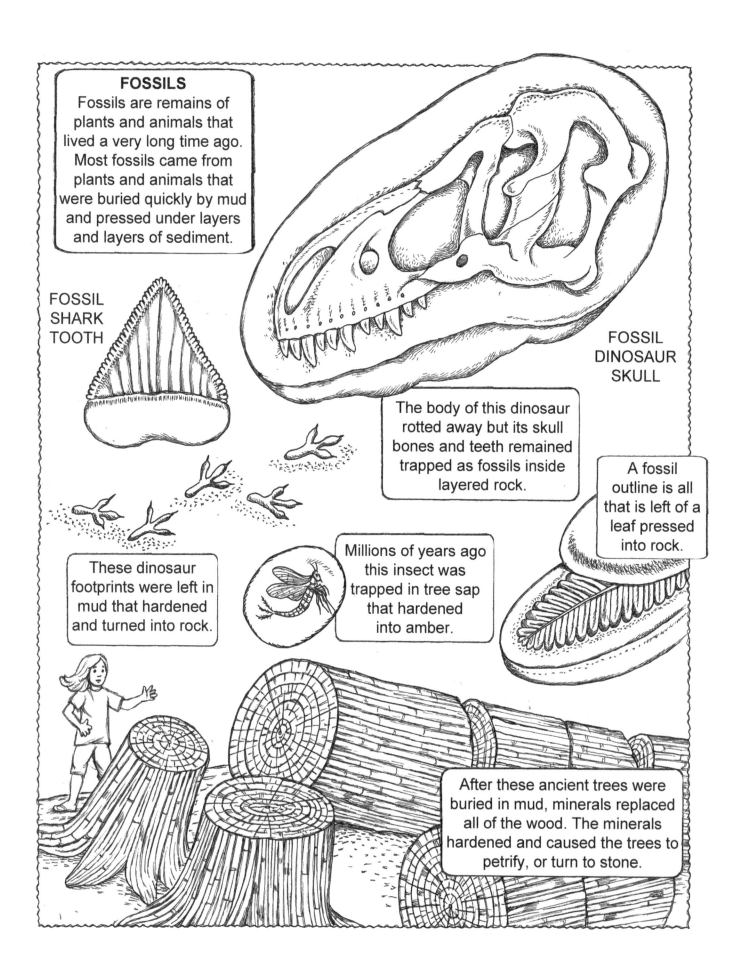

FOSSILS
Fossils are remains of plants and animals that lived a very long time ago. Most fossils came from plants and animals that were buried quickly by mud and pressed under layers and layers of sediment.

FOSSIL SHARK TOOTH

FOSSIL DINOSAUR SKULL

The body of this dinosaur rotted away but its skull bones and teeth remained trapped as fossils inside layered rock.

A fossil outline is all that is left of a leaf pressed into rock.

These dinosaur footprints were left in mud that hardened and turned into rock.

Millions of years ago this insect was trapped in tree sap that hardened into amber.

After these ancient trees were buried in mud, minerals replaced all of the wood. The minerals hardened and caused the trees to petrify, or turn to stone.

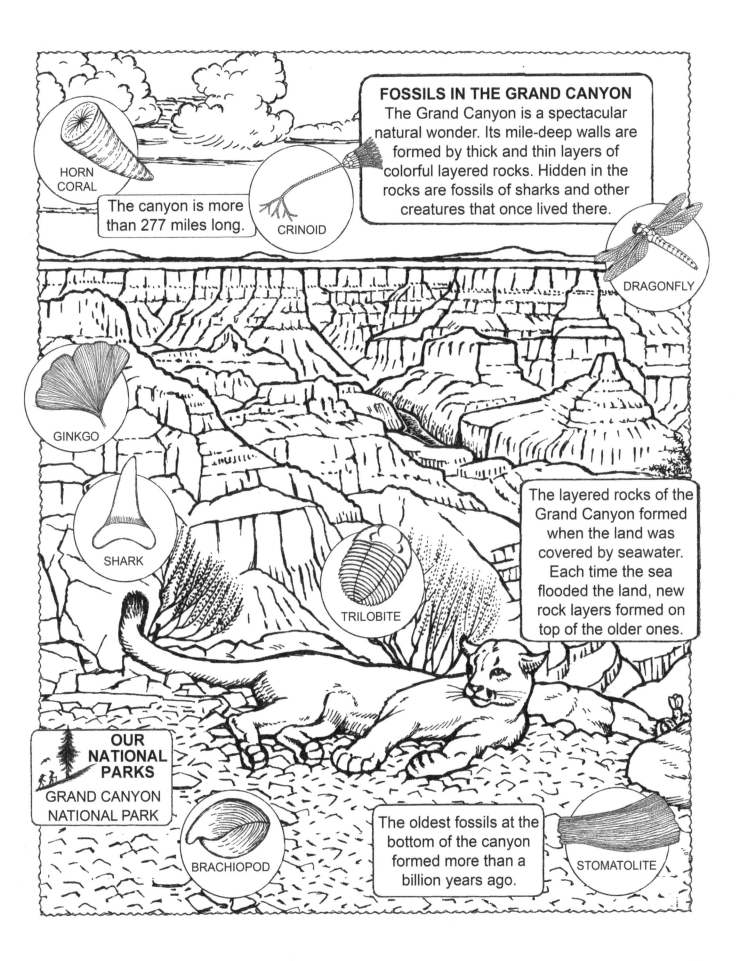

FOSSILS IN THE GRAND CANYON
The Grand Canyon is a spectacular natural wonder. Its mile-deep walls are formed by thick and thin layers of colorful layered rocks. Hidden in the rocks are fossils of sharks and other creatures that once lived there.

HORN CORAL

The canyon is more than 277 miles long.

CRINOID

DRAGONFLY

GINKGO

SHARK

TRILOBITE

The layered rocks of the Grand Canyon formed when the land was covered by seawater. Each time the sea flooded the land, new rock layers formed on top of the older ones.

OUR NATIONAL PARKS
GRAND CANYON NATIONAL PARK

BRACHIOPOD

The oldest fossils at the bottom of the canyon formed more than a billion years ago.

STOMATOLITE

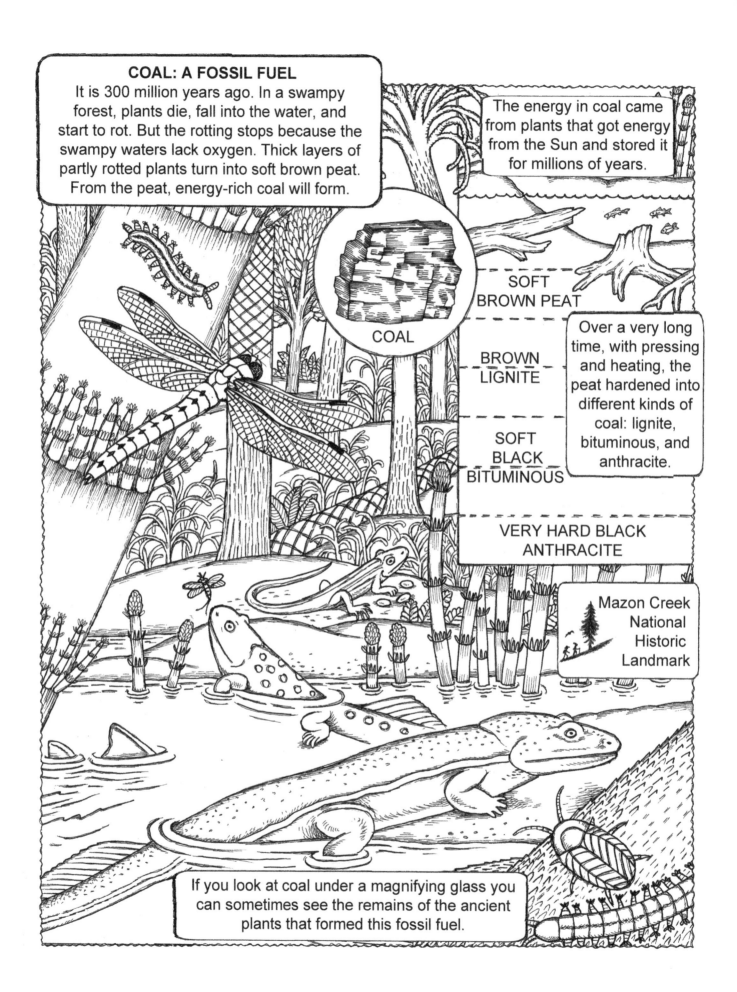

COAL: A FOSSIL FUEL
It is 300 million years ago. In a swampy forest, plants die, fall into the water, and start to rot. But the rotting stops because the swampy waters lack oxygen. Thick layers of partly rotted plants turn into soft brown peat. From the peat, energy-rich coal will form.

The energy in coal came from plants that got energy from the Sun and stored it for millions of years.

COAL

SOFT BROWN PEAT

BROWN LIGNITE

SOFT BLACK BITUMINOUS

Over a very long time, with pressing and heating, the peat hardened into different kinds of coal: lignite, bituminous, and anthracite.

VERY HARD BLACK ANTHRACITE

Mazon Creek National Historic Landmark

If you look at coal under a magnifying glass you can sometimes see the remains of the ancient plants that formed this fossil fuel.

18

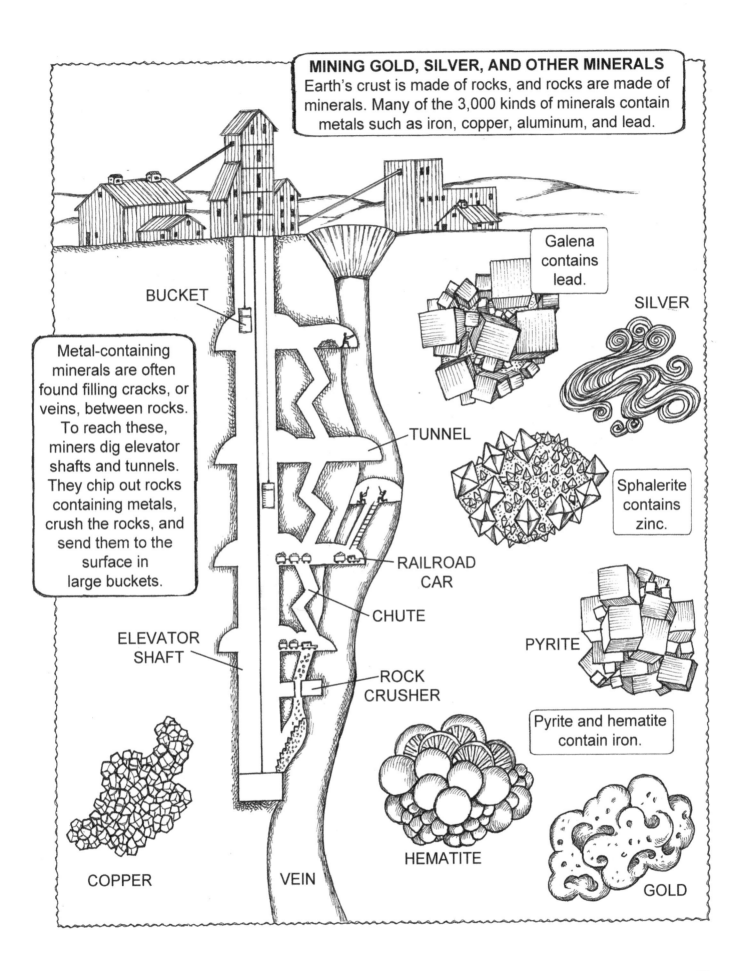

MINING GOLD, SILVER, AND OTHER MINERALS
Earth's crust is made of rocks, and rocks are made of minerals. Many of the 3,000 kinds of minerals contain metals such as iron, copper, aluminum, and lead.

BUCKET

Galena contains lead.

SILVER

Metal-containing minerals are often found filling cracks, or veins, between rocks. To reach these, miners dig elevator shafts and tunnels. They chip out rocks containing metals, crush the rocks, and send them to the surface in large buckets.

TUNNEL

Sphalerite contains zinc.

RAILROAD CAR

CHUTE

PYRITE

ELEVATOR SHAFT

ROCK CRUSHER

Pyrite and hematite contain iron.

COPPER

VEIN

HEMATITE

GOLD

19

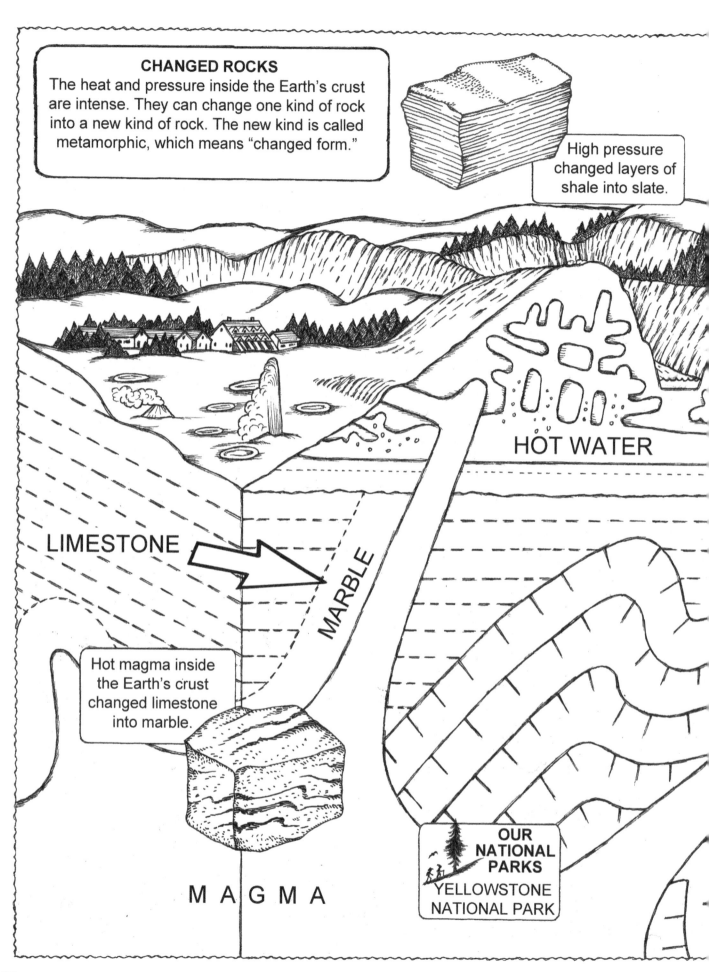

CHANGED ROCKS
The heat and pressure inside the Earth's crust are intense. They can change one kind of rock into a new kind of rock. The new kind is called metamorphic, which means "changed form."

High pressure changed layers of shale into slate.

HOT WATER

LIMESTONE

MARBLE

Hot magma inside the Earth's crust changed limestone into marble.

MAGMA

OUR NATIONAL PARKS
YELLOWSTONE NATIONAL PARK

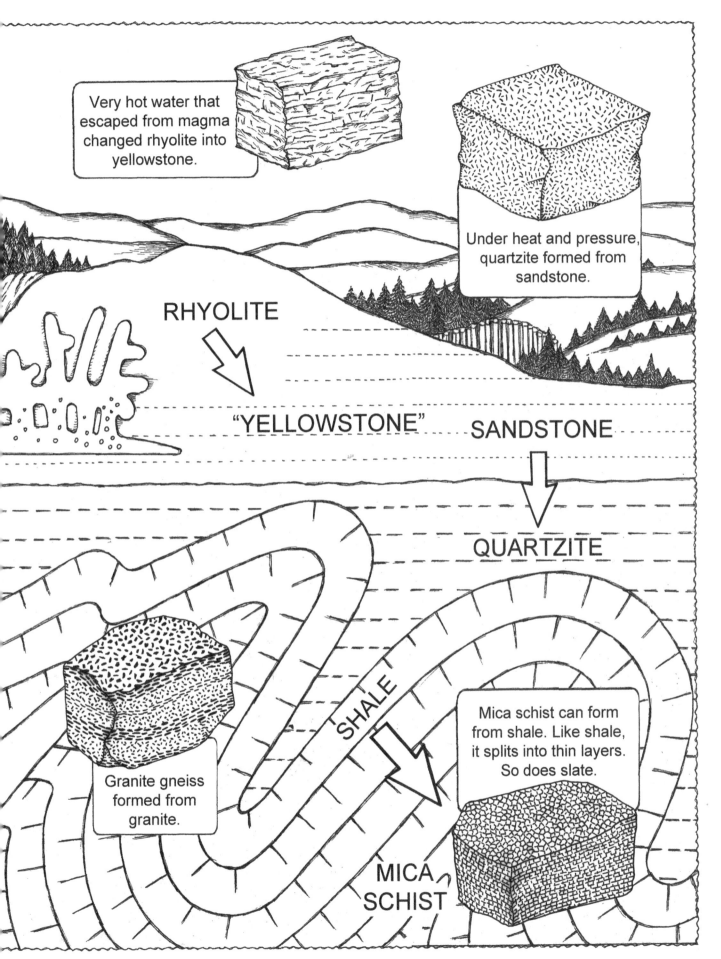

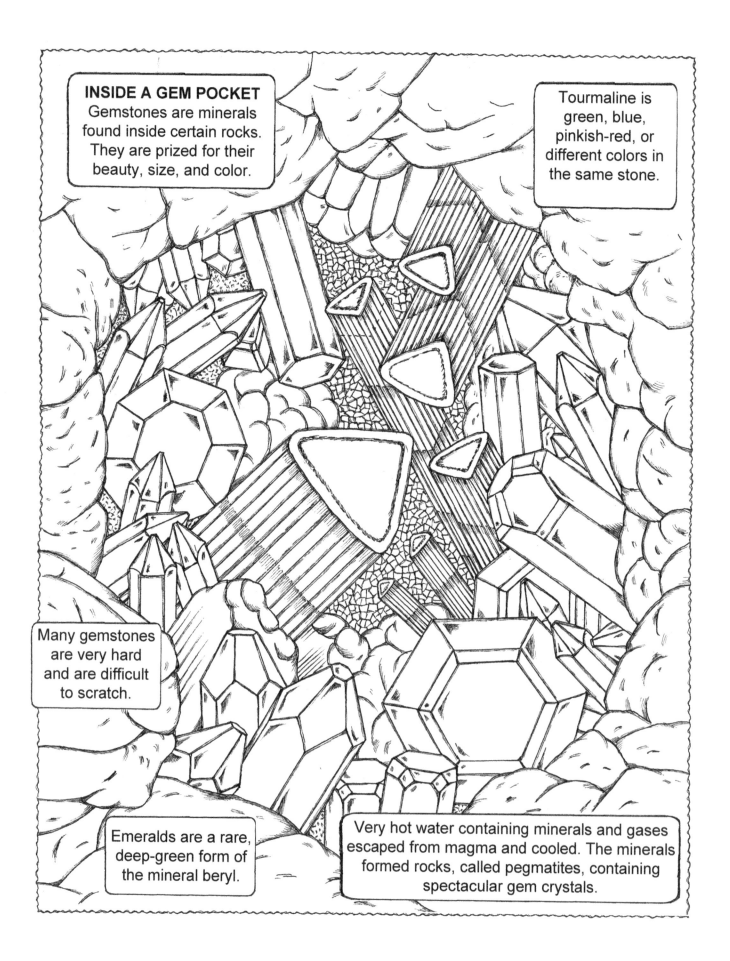

INSIDE A GEM POCKET
Gemstones are minerals found inside certain rocks. They are prized for their beauty, size, and color.

Tourmaline is green, blue, pinkish-red, or different colors in the same stone.

Many gemstones are very hard and are difficult to scratch.

Emeralds are a rare, deep-green form of the mineral beryl.

Very hot water containing minerals and gases escaped from magma and cooled. The minerals formed rocks, called pegmatites, containing spectacular gem crystals.

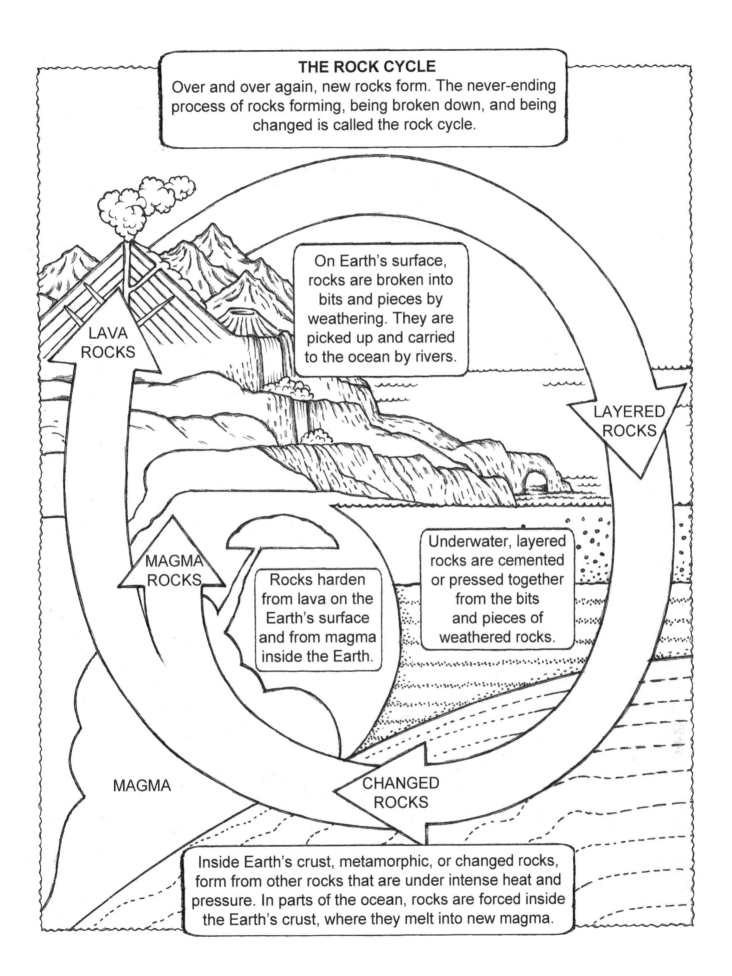

THE ROCK CYCLE
Over and over again, new rocks form. The never-ending process of rocks forming, being broken down, and being changed is called the rock cycle.

LAVA ROCKS

On Earth's surface, rocks are broken into bits and pieces by weathering. They are picked up and carried to the ocean by rivers.

LAYERED ROCKS

MAGMA ROCKS

Rocks harden from lava on the Earth's surface and from magma inside the Earth.

Underwater, layered rocks are cemented or pressed together from the bits and pieces of weathered rocks.

MAGMA

CHANGED ROCKS

Inside Earth's crust, metamorphic, or changed rocks, form from other rocks that are under intense heat and pressure. In parts of the ocean, rocks are forced inside the Earth's crust, where they melt into new magma.

THE WATER CYCLE
Nothing on Earth is like water. As rivers, it carves out canyons. As ice, it sharpens mountain peaks. It wears down rocks and carries pieces of weathered rocks from the land to the ocean.

Water vapor rises with the air, cools, and condenses – turning back into tiny water drops that form clouds.

As the Sun heats the ocean, some water evaporates, turning into the gas water vapor.

As the air in clouds cools, water drops form and fall as rain or snow.

Rain and melted snow flow into rivers or are soaked up by the ground.

Rivers return water back to the ocean.

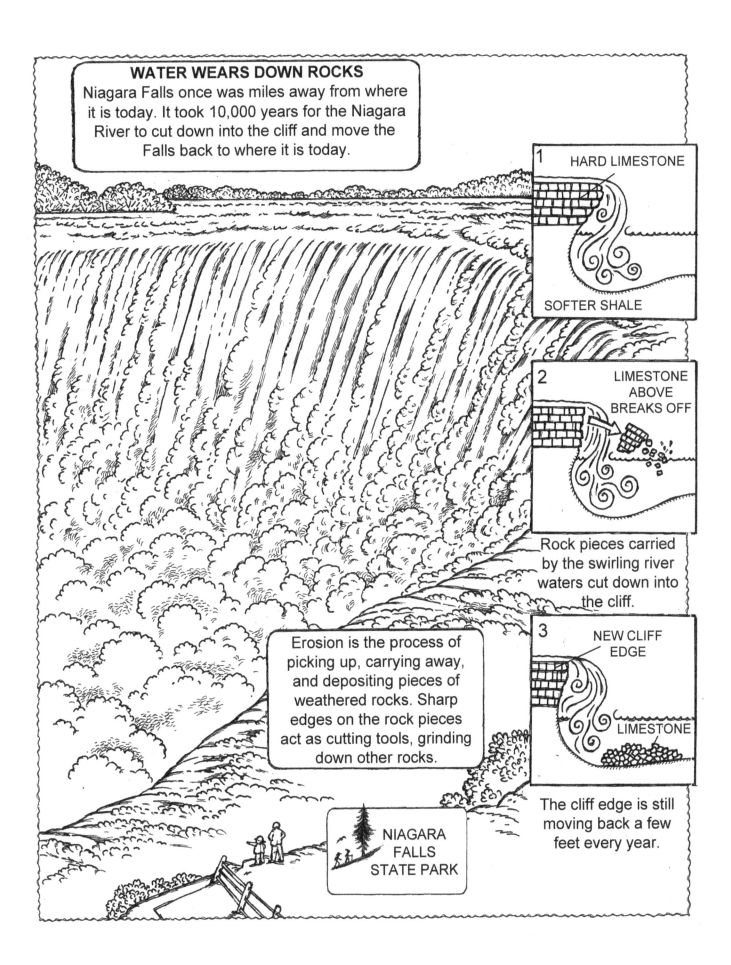

WATER WEARS DOWN ROCKS
Niagara Falls once was miles away from where it is today. It took 10,000 years for the Niagara River to cut down into the cliff and move the Falls back to where it is today.

1 HARD LIMESTONE

SOFTER SHALE

2 LIMESTONE ABOVE BREAKS OFF

Rock pieces carried by the swirling river waters cut down into the cliff.

Erosion is the process of picking up, carrying away, and depositing pieces of weathered rocks. Sharp edges on the rock pieces act as cutting tools, grinding down other rocks.

3 NEW CLIFF EDGE

LIMESTONE

The cliff edge is still moving back a few feet every year.

NIAGARA FALLS STATE PARK

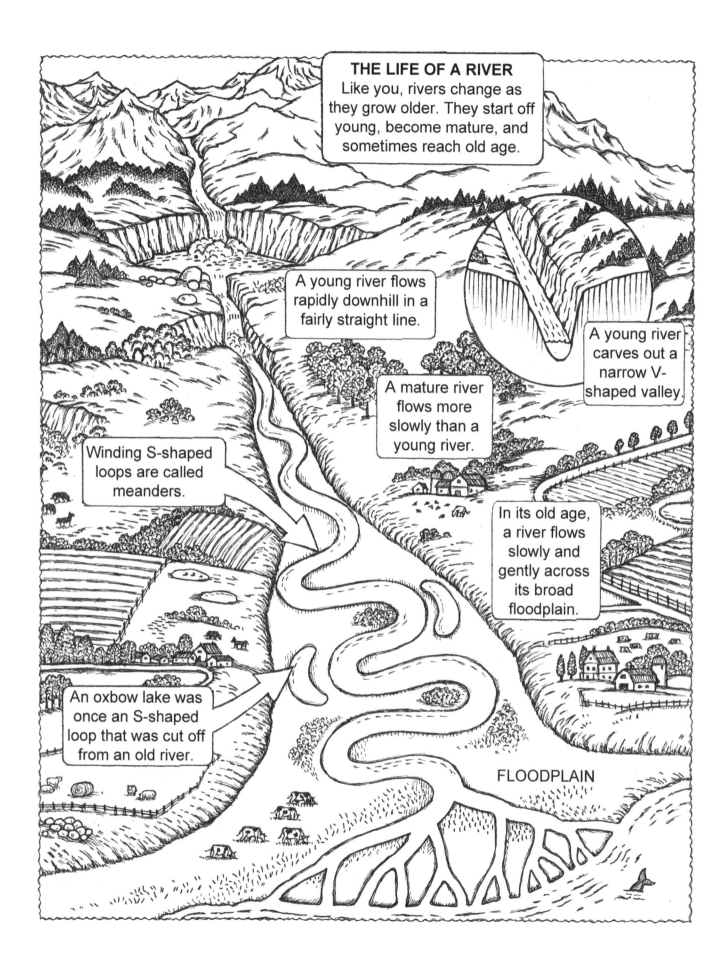

THE LIFE OF A RIVER
Like you, rivers change as they grow older. They start off young, become mature, and sometimes reach old age.

A young river flows rapidly downhill in a fairly straight line.

A young river carves out a narrow V-shaped valley.

A mature river flows more slowly than a young river.

Winding S-shaped loops are called meanders.

In its old age, a river flows slowly and gently across its broad floodplain.

An oxbow lake was once an S-shaped loop that was cut off from an old river.

FLOODPLAIN

THE OCEAN SHAPES THE LAND

When ocean waves pound into cliffs, they break off pieces of rock. The constant pounding can wear away large holes and form a sea arch.

OUR NATIONAL PARKS
ACADIA NATIONAL PARK

Waves pound against a cliff.

They carve a sea cave out of the rock.

More pounding cuts a hole to form a sea arch.

Over time, the crashing waves wear away the arch until it collapses.

WIND AND SAND
Deserts are the driest places on Earth. Little rain falls, and few plants grow there. Winds that sweep across deserts carry away loose soil, lowering the land layer by layer. It takes strong winds to lift and blast sand into boulders and cliffs.

OUR NATIONAL PARKS
ARCHES NATIONAL PARK

A river once flowed here. It wore a hole through rock, forming a natural bridge. Today wind erodes the bridge, changing its shape.

When winds slow down, they drop sand in mounds, called dunes.

BLOWING SAND

Very strong winds can lift sand only a few feet off the ground, but windblown sand can change the shape of rocks.

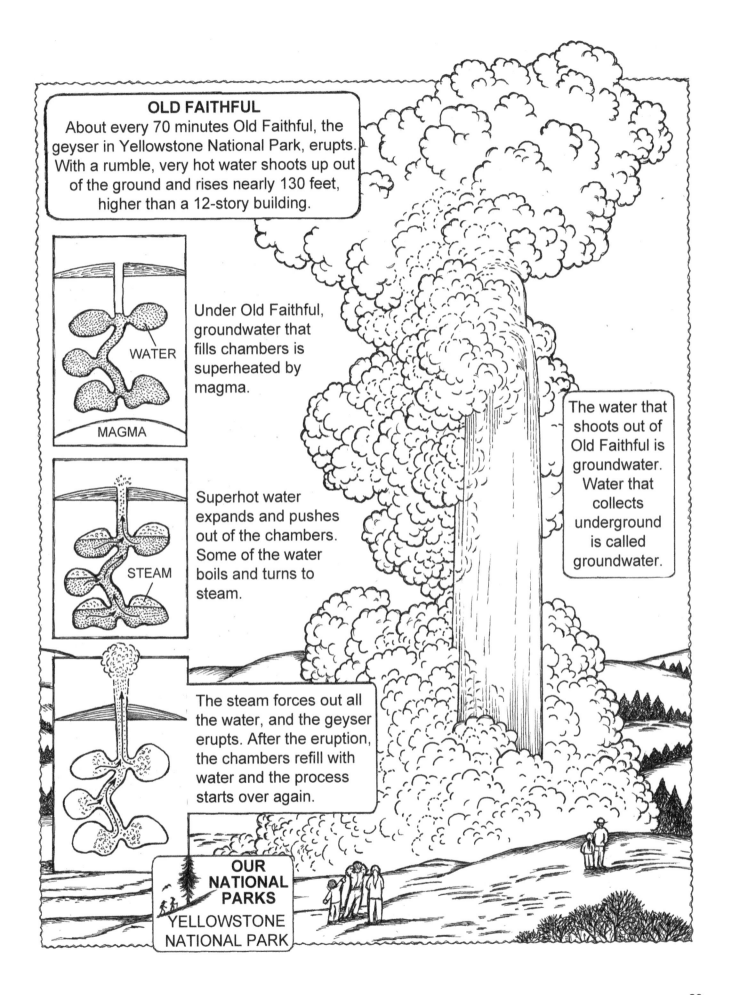

OLD FAITHFUL
About every 70 minutes Old Faithful, the geyser in Yellowstone National Park, erupts. With a rumble, very hot water shoots up out of the ground and rises nearly 130 feet, higher than a 12-story building.

Under Old Faithful, groundwater that fills chambers is superheated by magma.

WATER

MAGMA

Superhot water expands and pushes out of the chambers. Some of the water boils and turns to steam.

STEAM

The water that shoots out of Old Faithful is groundwater. Water that collects underground is called groundwater.

The steam forces out all the water, and the geyser erupts. After the eruption, the chambers refill with water and the process starts over again.

OUR NATIONAL PARKS
YELLOWSTONE NATIONAL PARK

29

CAVES
More than 100 years ago, cowboys in New Mexico climbed into a deep pit in the ground and came across tunnels and an immense underground chamber. They discovered the vast limestone cave called Carlsbad Caverns.

The Big Room in Carlsbad Caverns is as tall as the United States Capitol Building.

NEVER explore a cave on your own. Explore with a guide who knows the cave.

OUR NATIONAL PARKS
CARLSBAD CAVERNS NATIONAL PARK

Groundwater containing small amounts of the gas carbon dioxide can seep into cracks in limestone. Slowly, the water eats away the rock. Over time, the cracks grow larger and larger to form a cave.

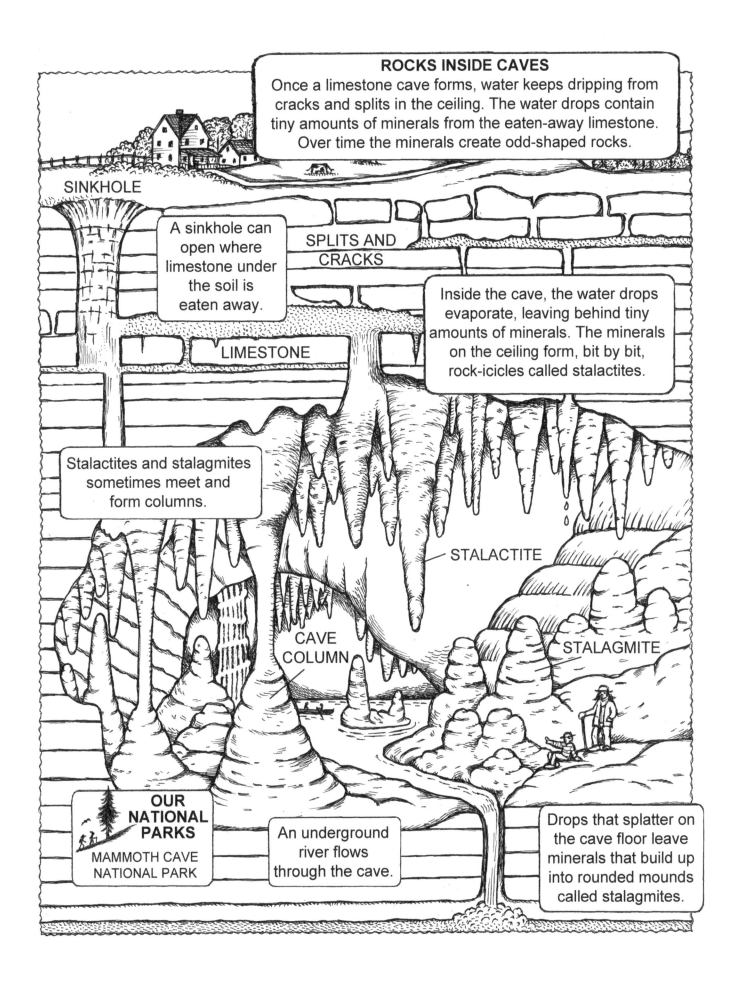

ROCKS INSIDE CAVES
Once a limestone cave forms, water keeps dripping from cracks and splits in the ceiling. The water drops contain tiny amounts of minerals from the eaten-away limestone. Over time the minerals create odd-shaped rocks.

SINKHOLE

A sinkhole can open where limestone under the soil is eaten away.

SPLITS AND CRACKS

Inside the cave, the water drops evaporate, leaving behind tiny amounts of minerals. The minerals on the ceiling form, bit by bit, rock-icicles called stalactites.

LIMESTONE

Stalactites and stalagmites sometimes meet and form columns.

STALACTITE

CAVE COLUMN

STALAGMITE

OUR NATIONAL PARKS

MAMMOTH CAVE NATIONAL PARK

An underground river flows through the cave.

Drops that splatter on the cave floor leave minerals that build up into rounded mounds called stalagmites.

31

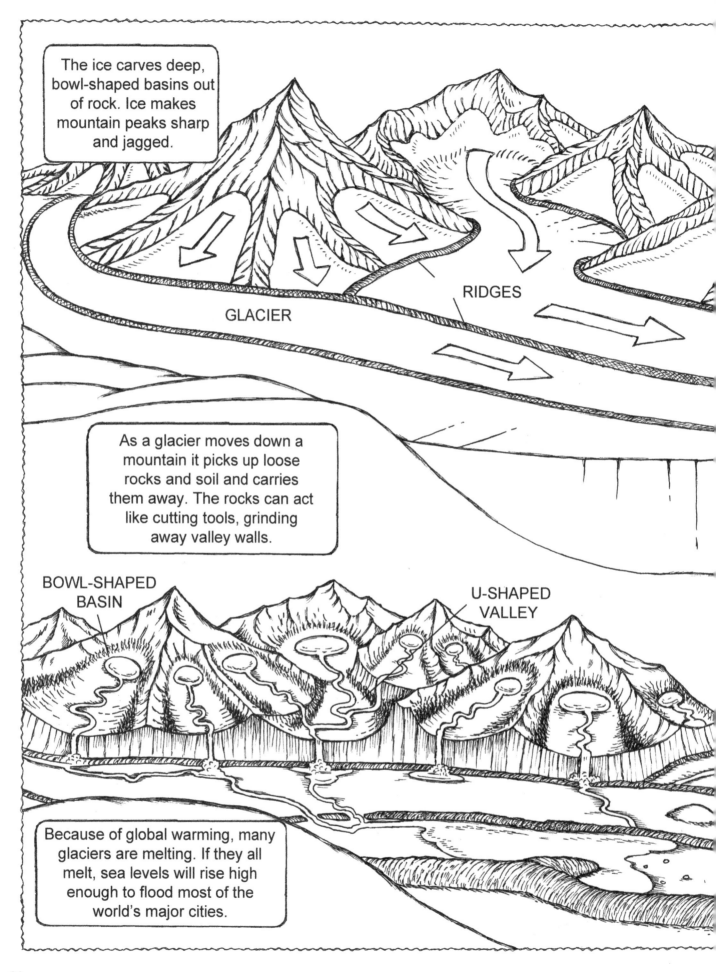

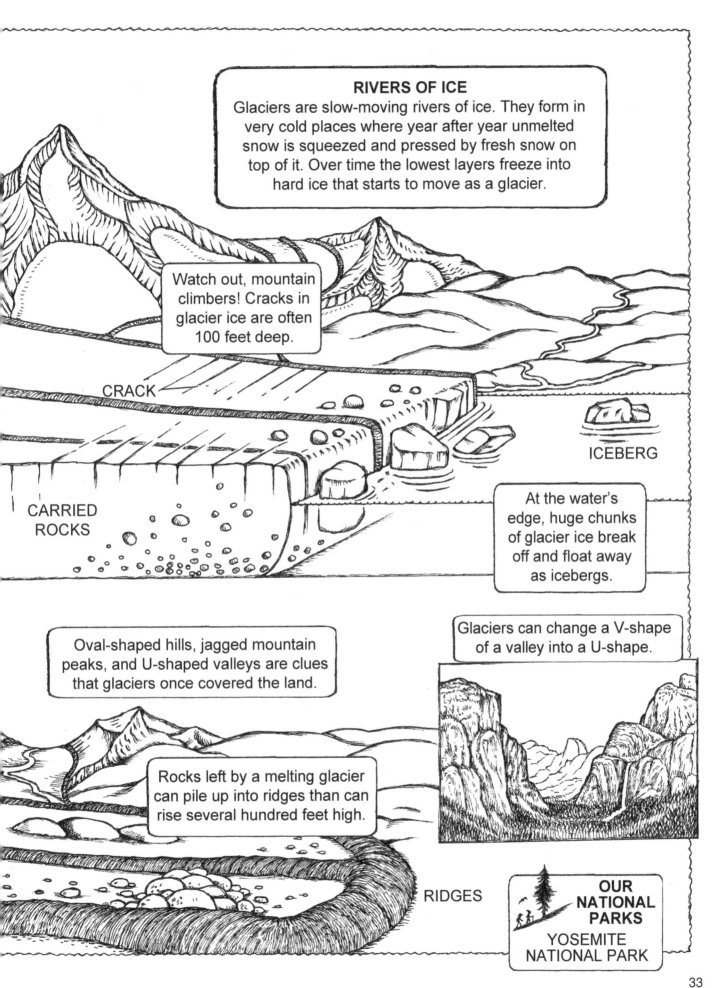

RIVERS OF ICE
Glaciers are slow-moving rivers of ice. They form in very cold places where year after year unmelted snow is squeezed and pressed by fresh snow on top of it. Over time the lowest layers freeze into hard ice that starts to move as a glacier.

Watch out, mountain climbers! Cracks in glacier ice are often 100 feet deep.

CRACK

CARRIED ROCKS

ICEBERG

At the water's edge, huge chunks of glacier ice break off and float away as icebergs.

Glaciers can change a V-shape of a valley into a U-shape.

Oval-shaped hills, jagged mountain peaks, and U-shaped valleys are clues that glaciers once covered the land.

Rocks left by a melting glacier can pile up into ridges than can rise several hundred feet high.

RIDGES

OUR NATIONAL PARKS
YOSEMITE NATIONAL PARK

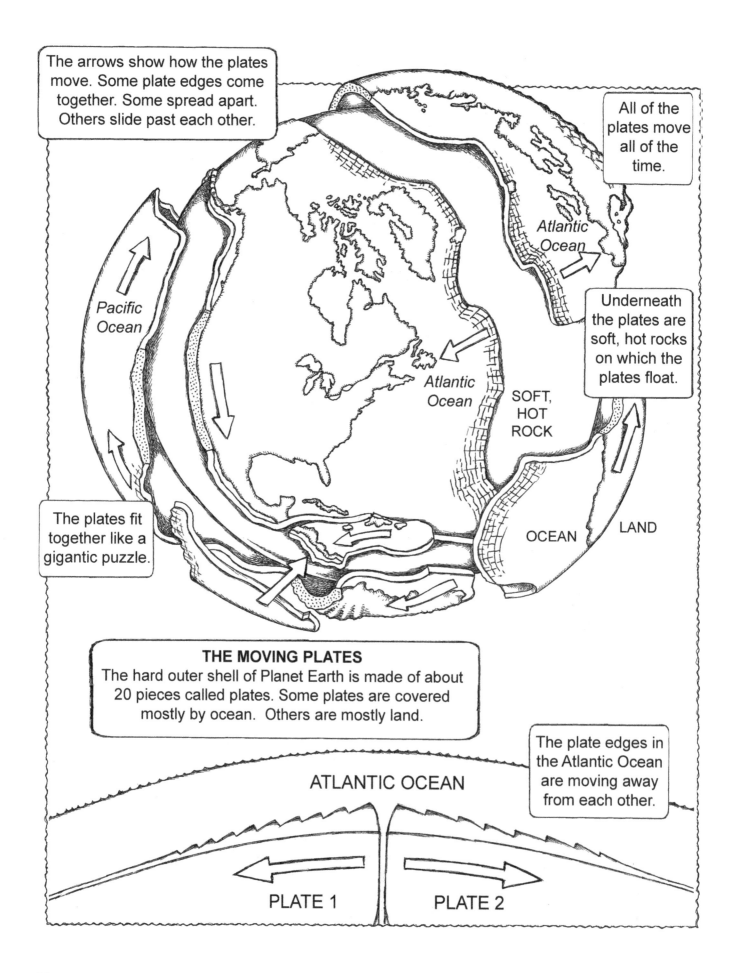

The arrows show how the plates move. Some plate edges come together. Some spread apart. Others slide past each other.

All of the plates move all of the time.

Pacific Ocean

Atlantic Ocean

Underneath the plates are soft, hot rocks on which the plates float.

Atlantic Ocean

SOFT, HOT ROCK

The plates fit together like a gigantic puzzle.

OCEAN   LAND

THE MOVING PLATES
The hard outer shell of Planet Earth is made of about 20 pieces called plates. Some plates are covered mostly by ocean. Others are mostly land.

The plate edges in the Atlantic Ocean are moving away from each other.

ATLANTIC OCEAN

PLATE 1   PLATE 2

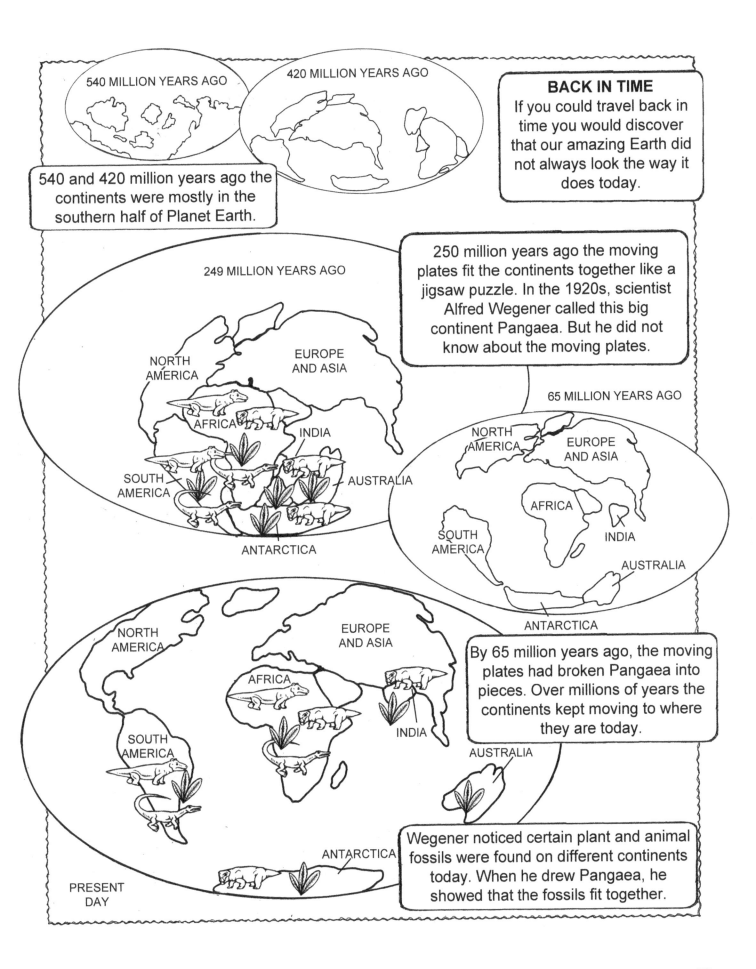

540 MILLION YEARS AGO

420 MILLION YEARS AGO

BACK IN TIME
If you could travel back in time you would discover that our amazing Earth did not always look the way it does today.

540 and 420 million years ago the continents were mostly in the southern half of Planet Earth.

249 MILLION YEARS AGO

250 million years ago the moving plates fit the continents together like a jigsaw puzzle. In the 1920s, scientist Alfred Wegener called this big continent Pangaea. But he did not know about the moving plates.

NORTH AMERICA

EUROPE AND ASIA

AFRICA

INDIA

SOUTH AMERICA

AUSTRALIA

ANTARCTICA

65 MILLION YEARS AGO

NORTH AMERICA

EUROPE AND ASIA

AFRICA

INDIA

SOUTH AMERICA

AUSTRALIA

ANTARCTICA

NORTH AMERICA

EUROPE AND ASIA

AFRICA

INDIA

SOUTH AMERICA

By 65 million years ago, the moving plates had broken Pangaea into pieces. Over millions of years the continents kept moving to where they are today.

AUSTRALIA

ANTARCTICA

PRESENT DAY

Wegener noticed certain plant and animal fossils were found on different continents today. When he drew Pangaea, he showed that the fossils fit together.

35

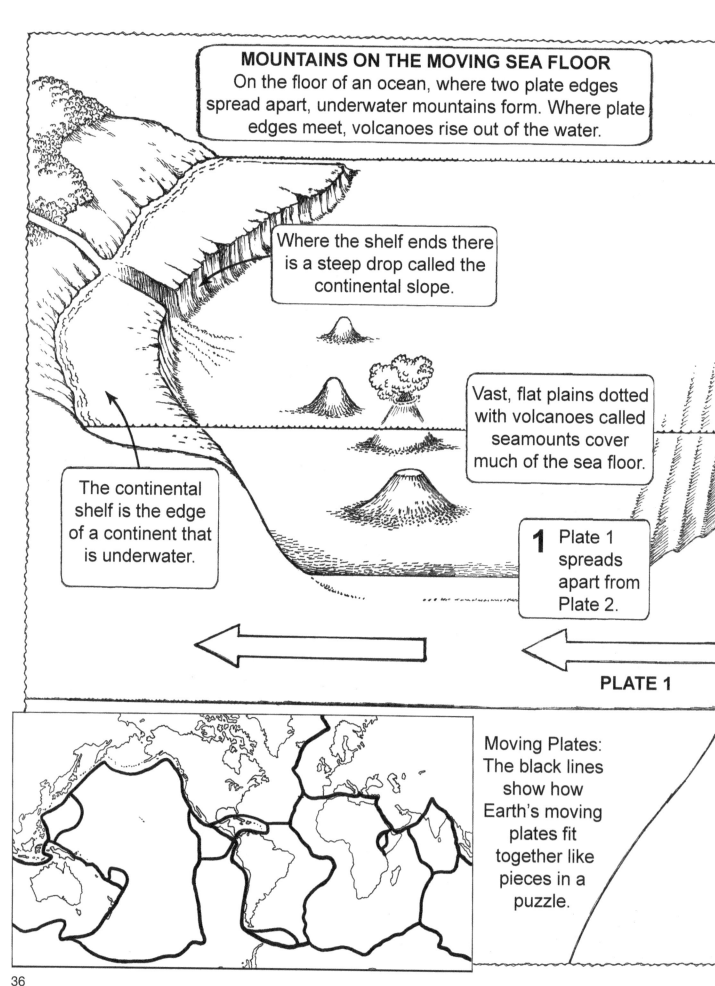

MOUNTAINS ON THE MOVING SEA FLOOR
On the floor of an ocean, where two plate edges spread apart, underwater mountains form. Where plate edges meet, volcanoes rise out of the water.

Where the shelf ends there is a steep drop called the continental slope.

Vast, flat plains dotted with volcanoes called seamounts cover much of the sea floor.

The continental shelf is the edge of a continent that is underwater.

1 Plate 1 spreads apart from Plate 2.

PLATE 1

Moving Plates: The black lines show how Earth's moving plates fit together like pieces in a puzzle.

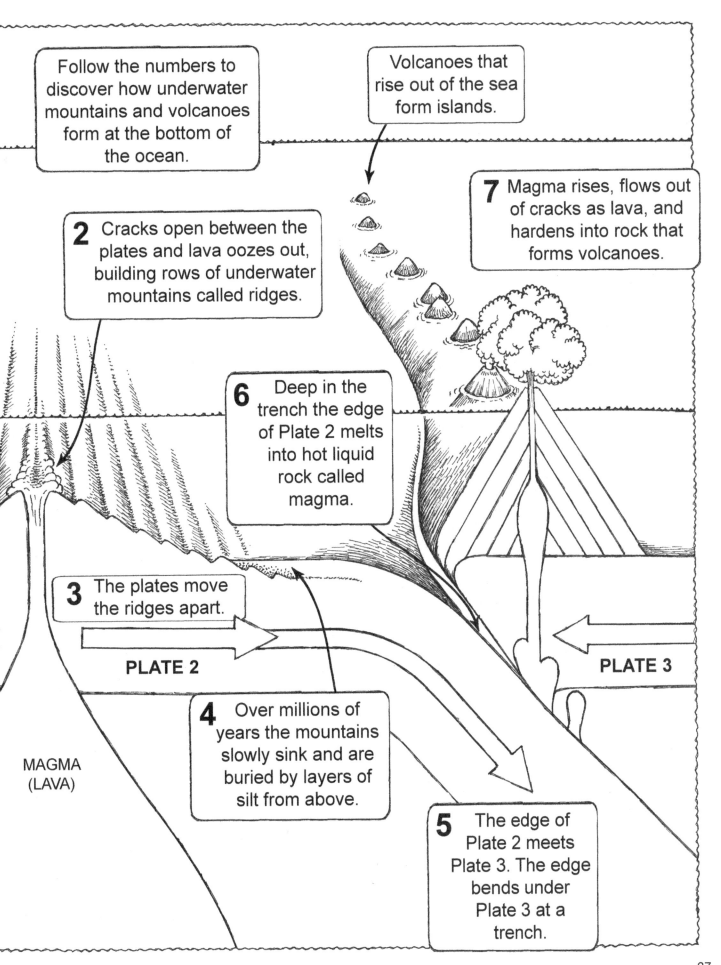

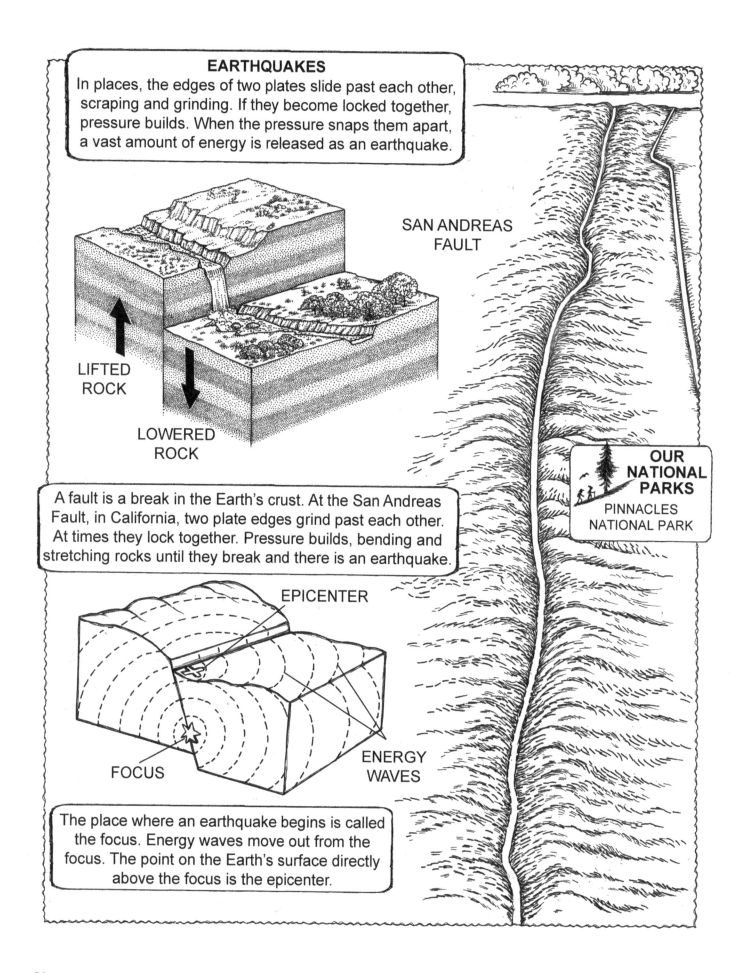

EARTHQUAKES
In places, the edges of two plates slide past each other, scraping and grinding. If they become locked together, pressure builds. When the pressure snaps them apart, a vast amount of energy is released as an earthquake.

SAN ANDREAS FAULT

LIFTED ROCK

LOWERED ROCK

A fault is a break in the Earth's crust. At the San Andreas Fault, in California, two plate edges grind past each other. At times they lock together. Pressure builds, bending and stretching rocks until they break and there is an earthquake.

EPICENTER

FOCUS

ENERGY WAVES

OUR NATIONAL PARKS
PINNACLES NATIONAL PARK

The place where an earthquake begins is called the focus. Energy waves move out from the focus. The point on the Earth's surface directly above the focus is the epicenter.

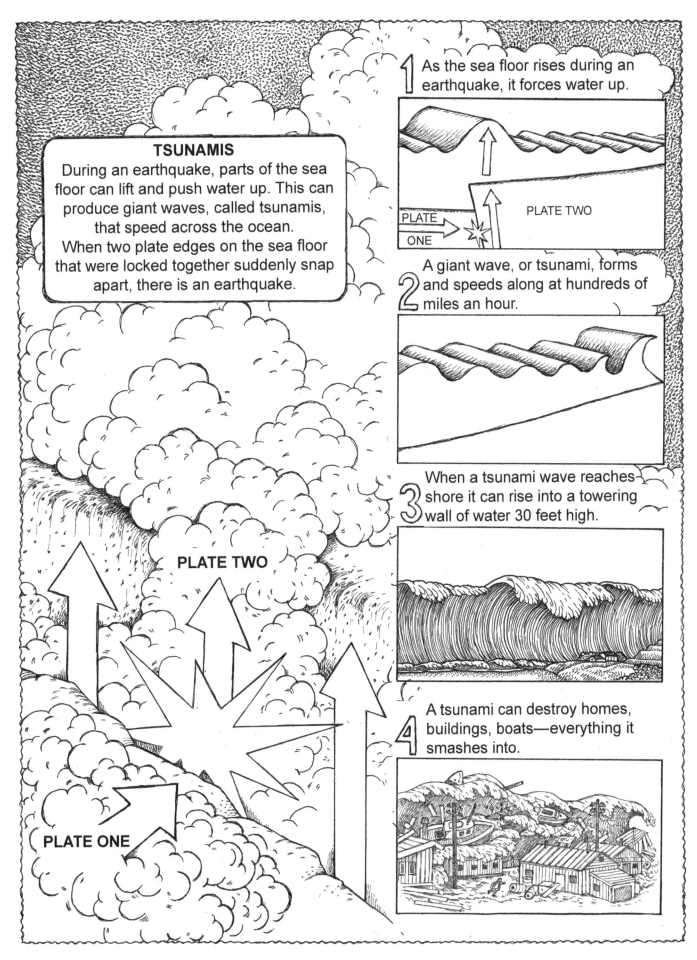

TSUNAMIS

During an earthquake, parts of the sea floor can lift and push water up. This can produce giant waves, called tsunamis, that speed across the ocean.

When two plate edges on the sea floor that were locked together suddenly snap apart, there is an earthquake.

PLATE TWO

PLATE ONE

1 As the sea floor rises during an earthquake, it forces water up.

PLATE ONE   PLATE TWO

2 A giant wave, or tsunami, forms and speeds along at hundreds of miles an hour.

3 When a tsunami wave reaches shore it can rise into a towering wall of water 30 feet high.

4 A tsunami can destroy homes, buildings, boats—everything it smashes into.

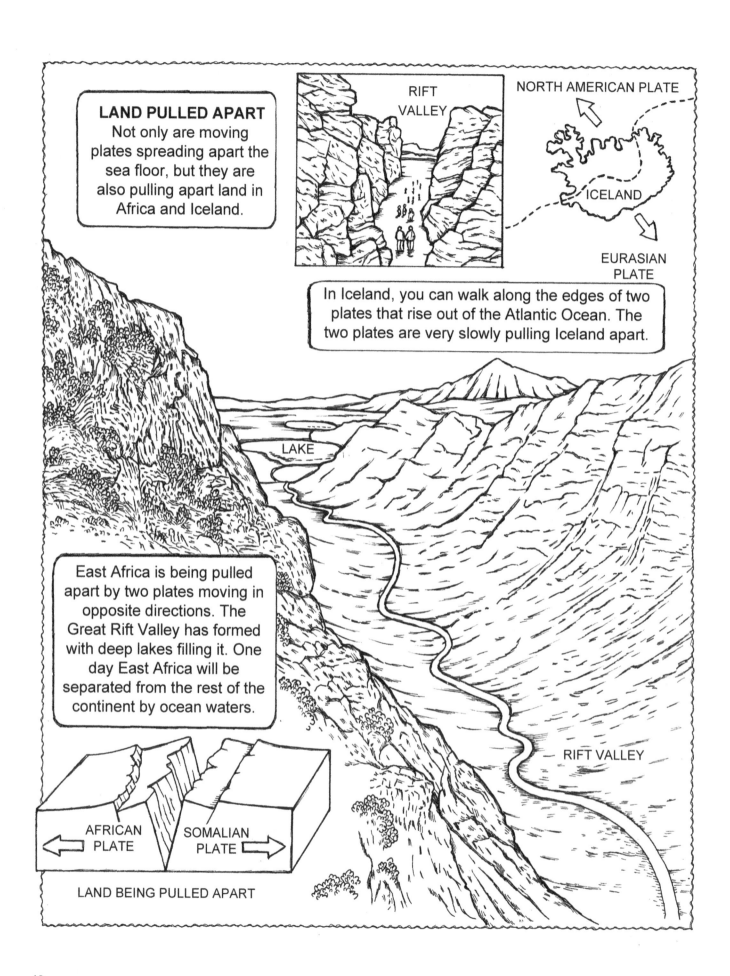

LAND PULLED APART
Not only are moving plates spreading apart the sea floor, but they are also pulling apart land in Africa and Iceland.

RIFT VALLEY

NORTH AMERICAN PLATE

ICELAND

EURASIAN PLATE

In Iceland, you can walk along the edges of two plates that rise out of the Atlantic Ocean. The two plates are very slowly pulling Iceland apart.

LAKE

East Africa is being pulled apart by two plates moving in opposite directions. The Great Rift Valley has formed with deep lakes filling it. One day East Africa will be separated from the rest of the continent by ocean waters.

RIFT VALLEY

AFRICAN PLATE

SOMALIAN PLATE

LAND BEING PULLED APART

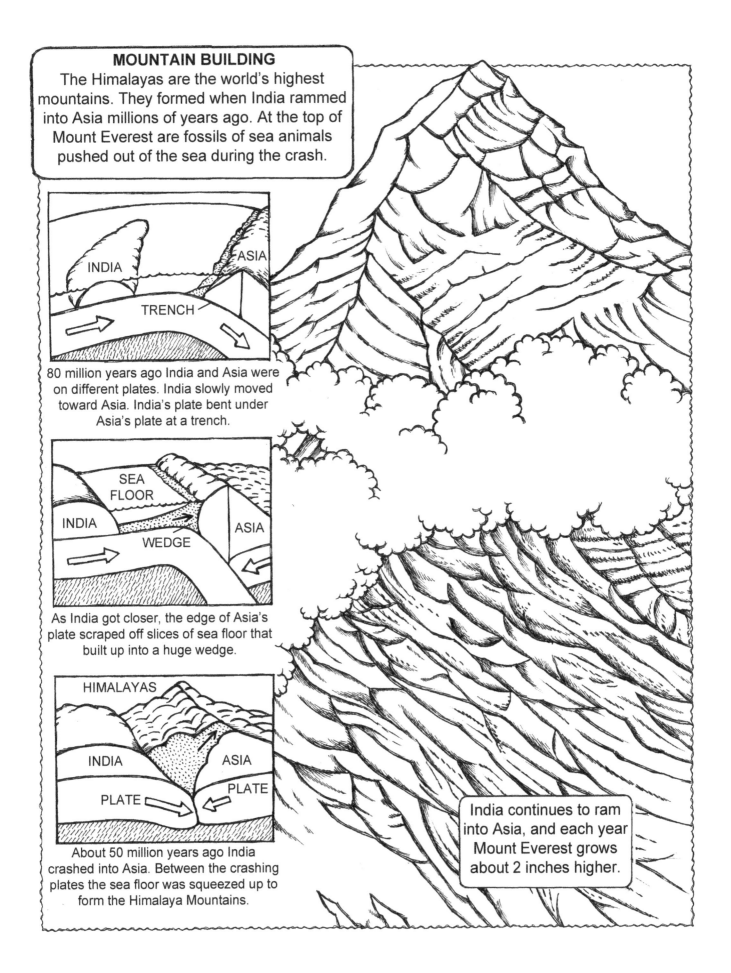

MOUNTAIN BUILDING
The Himalayas are the world's highest mountains. They formed when India rammed into Asia millions of years ago. At the top of Mount Everest are fossils of sea animals pushed out of the sea during the crash.

INDIA

ASIA

TRENCH

80 million years ago India and Asia were on different plates. India slowly moved toward Asia. India's plate bent under Asia's plate at a trench.

SEA FLOOR

INDIA

ASIA

WEDGE

As India got closer, the edge of Asia's plate scraped off slices of sea floor that built up into a huge wedge.

HIMALAYAS

INDIA

ASIA

PLATE

PLATE

About 50 million years ago India crashed into Asia. Between the crashing plates the sea floor was squeezed up to form the Himalaya Mountains.

India continues to ram into Asia, and each year Mount Everest grows about 2 inches higher.

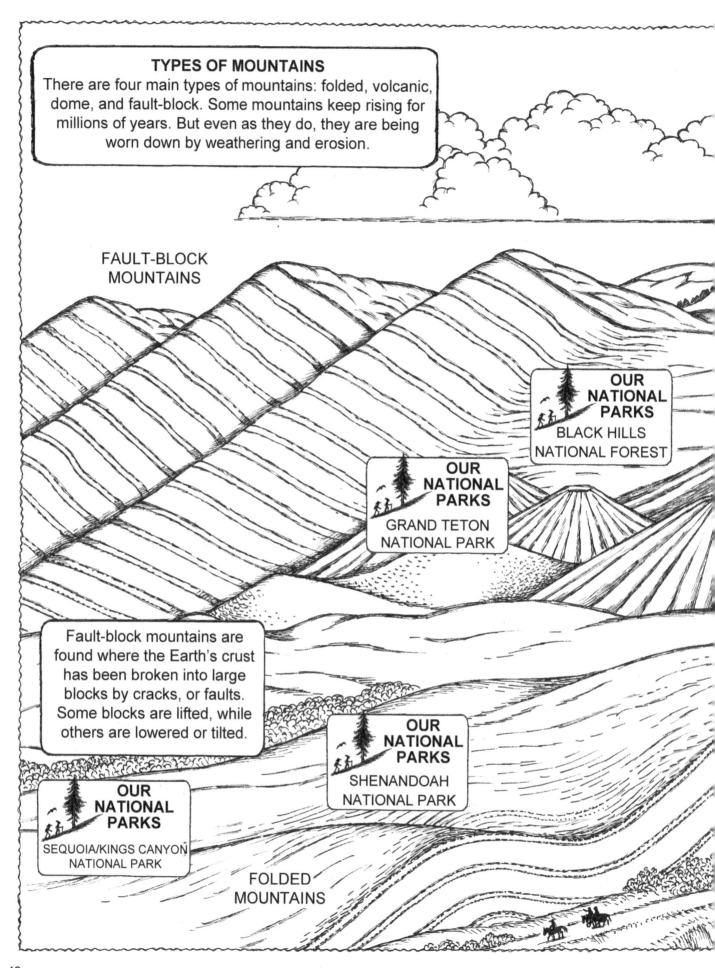

TYPES OF MOUNTAINS
There are four main types of mountains: folded, volcanic, dome, and fault-block. Some mountains keep rising for millions of years. But even as they do, they are being worn down by weathering and erosion.

FAULT-BLOCK
MOUNTAINS

OUR
NATIONAL
PARKS
BLACK HILLS
NATIONAL FOREST

OUR
NATIONAL
PARKS
GRAND TETON
NATIONAL PARK

Fault-block mountains are found where the Earth's crust has been broken into large blocks by cracks, or faults. Some blocks are lifted, while others are lowered or tilted.

OUR
NATIONAL
PARKS
SHENANDOAH
NATIONAL PARK

OUR
NATIONAL
PARKS
SEQUOIA/KINGS CANYON
NATIONAL PARK

FOLDED
MOUNTAINS

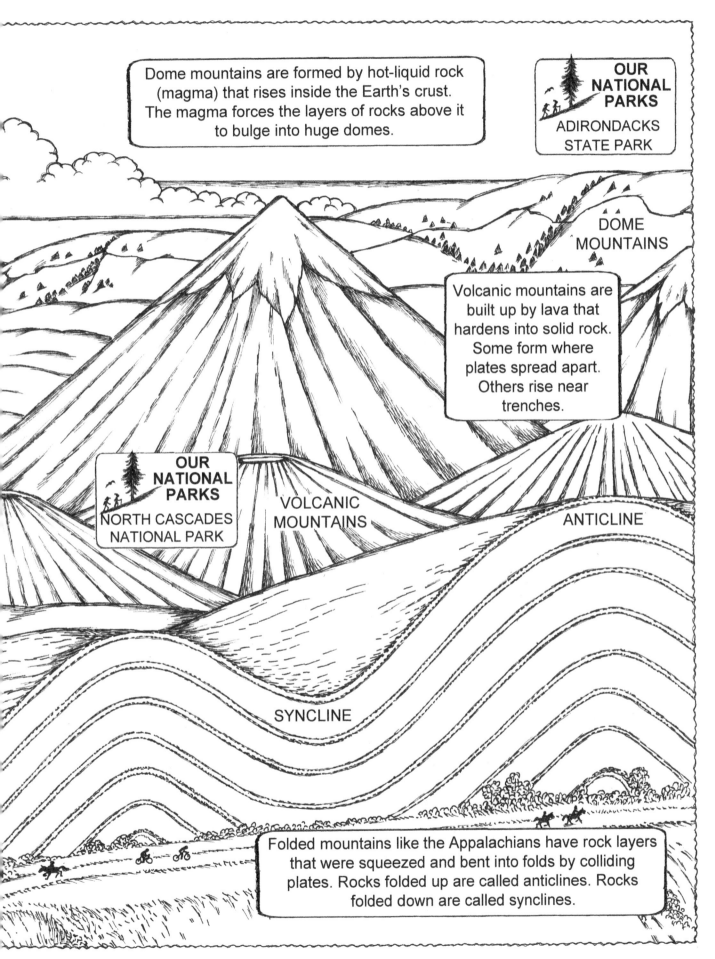

Dome mountains are formed by hot-liquid rock (magma) that rises inside the Earth's crust. The magma forces the layers of rocks above it to bulge into huge domes.

OUR NATIONAL PARKS
ADIRONDACKS STATE PARK

DOME MOUNTAINS

Volcanic mountains are built up by lava that hardens into solid rock. Some form where plates spread apart. Others rise near trenches.

OUR NATIONAL PARKS
NORTH CASCADES NATIONAL PARK

VOLCANIC MOUNTAINS

ANTICLINE

SYNCLINE

Folded mountains like the Appalachians have rock layers that were squeezed and bent into folds by colliding plates. Rocks folded up are called anticlines. Rocks folded down are called synclines.

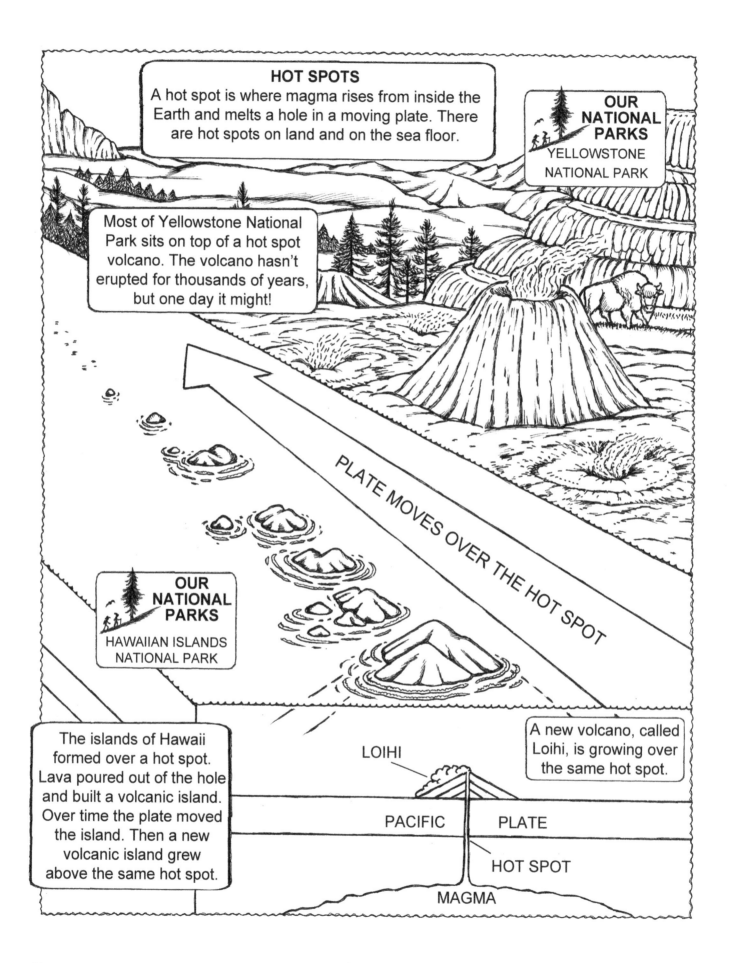

ATMOSPHERE

Every year, millions of chunks of space rocks and metals, called meteors, burn up in Earth's atmosphere. Those that reach Earth's surface are called meteorites.

ROCKS FROM SPACE
Scientists think that about 65 million years ago a giant rock, called an asteroid, crashed into Earth. The crash blasted tons of rock into the air, which filled with dust and smoke from fires that blocked sunlight for months.

METEOR

Without sunlight, plants could not make food for animals. The crash led to the end of the dinosaurs and about 75% of all other animals and plants.

ASTEROID

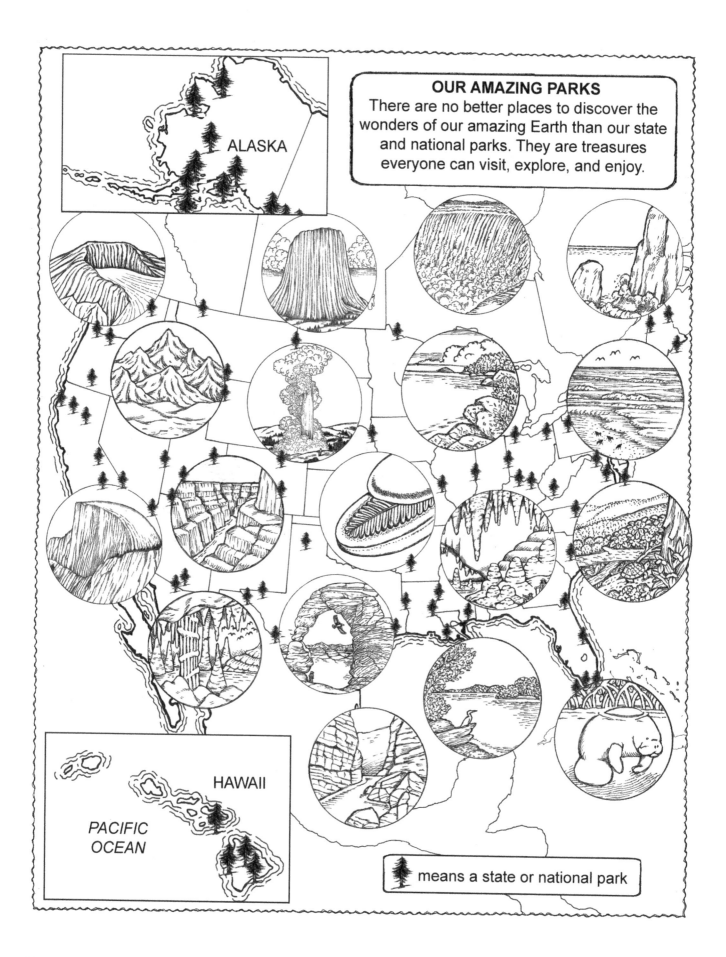

OUR AMAZING PARKS
There are no better places to discover the wonders of our amazing Earth than our state and national parks. They are treasures everyone can visit, explore, and enjoy.

ALASKA

HAWAII

PACIFIC OCEAN

🌲 means a state or national park